MAKE AN EGG FLOAT

Written by
Read With You Science Academy

Read With You

The characters and events portrayed in this book are fictitious. Any similarity to real persons, living or dead, is coincidental and not intended by the author.

Published by Read With You Publishing
Designed by Read With You Science Academy
Read With You and associated logos are trademarks and/or registered trademarks of Read With You L.L.C.

ISBN-13: 978-1-64616-109-6
ISBN-10: 1-64616-109-2

Third Edition April 2021
Printed in the United States of America.

Welcome to the

READ WITH YOU SCIENCE ACADEMY!

Today we will do an experiment called

MAKE AN EGG FLOAT

Before we begin, let's review some important words:

dense:
Having parts that are close together.

liquid:
Something that flows freely like water; not a solid or a gas.

float:
To rest on top of a liquid.

QUESTION:
What I want to know.

Will an egg float in water?

☐ Yes, an egg will float in water.

☐ No, an egg will not float in water.

Remember:
Be a careful scientist! Clean up any water that spills right away!

HYPOTHESIS:

What I think will happen and why.

I think an egg **will** float in water because...

I think an egg **won't** float in water because...

The egg **will** float because:	The egg **will not** float because:
The egg is lighter than the water.	The egg is too heavy to float in water.
A solid object will float in a liquid.	A solid object will not float in a liquid.

What do you think? Circle your hypothesis!

EXPERIMENT:
How we will test our hypothesis.

THINGS WE NEED

- 1 egg
- A tall, clear drinking glass or jar
- Room temperature water
- Table salt
- A clear bowl or pan (in case the water spills)
- Paper towels or a towel (for clean up!)

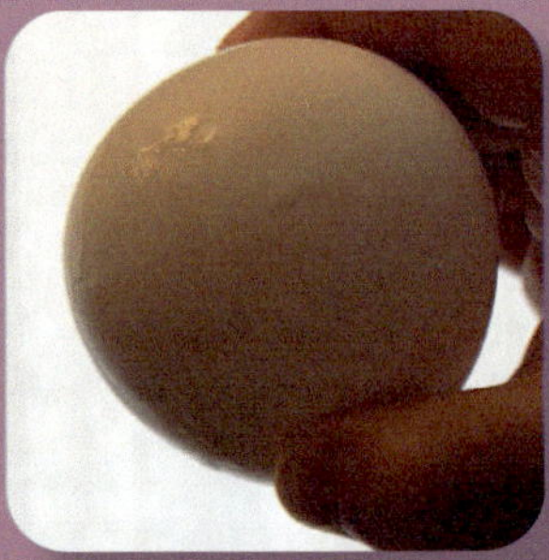
egg

clear glass

water

salt

pan

paper towels

PROCEDURE:

The steps to complete the experiment.

Step 1A

- Pour water into the glass until it is about three fourths (¾) full.

Step 2A

- Gently lower the egg into the water and **observe** what happens.
- Remember to put the glass over a bowl or in a pan in case your **liquid** spills!

- Remember: Use a towel or paper towel to clean any water on the table or floor!

Write down what you see

In your science journal,

write what you **observed!**

- What was your **hypothesis**? (Look on page 5 of this book!)
- Was your hypothesis correct?
- Why do you think this happened?

Let's do it again!

Do you think the egg will **float** in salt water? Let's do an experiment to find out. You will need water, a tall clear drinking glass, table salt, and an egg.

QUESTION:

What I want to know.

Will an egg float in salt water?

HYPOTHESIS:

What I think will happen and why.

I **think** the egg will float in the salt water because...

I **don't think** the egg will float in the salt water because...

HYPOTHESIS:

What I think will happen and why.

An egg **will** float in salt water because:	An egg **will not** float in salt water because:
Adding salt to the water makes it denser.	The egg is too heavy to float in water.
If the liquid is denser than the solid, the solid will float.	A solid object will not float in a liquid.

What do you think? Circle your hypothesis!

Write your new hypothesis in your science journal.

PROCEDURE:
The steps to complete the experiment.

Step 1B

Pour water into the drinking glass until about one half (½) of the glass is full.

Step 2B

- Stir in lots of salt (about 6 tablespoons).

Step 3B

- Carefully add plain water until the glass is nearly full.
- Be careful not to disturb or mix the salt water with the plain water.
- Remember to put the glass in a bowl or pan in case the water spills.

Step 4B

- Gently lower the egg into the water and observe what happens.
- Remember: Use a towel or paper towel to clean any water on the table or floor.

Write down what you see

In your science journal,

write your results!

- What was your hypothesis? (Look on page 12 of this book!)
- Was your hypothesis correct?
- Describe what happened to the egg.

Why did this work?

- This experiment worked because salt water is denser than ordinary water. It is easier for an object to float in a dense liquid.
- When you lowered the egg into the liquid, it dropped through the normal water until it reached the salt water. The salt water is dense enough for the egg to float. This is why the egg seemed to be floating in the middle of the glass.
- If this didn't work for you, try it again, but be careful not to mix the plain and salt water!

SCIENCE

- This experiment demonstrated how an object's **density** affects its **buoyancy**. Density is a measurement that compares the amount of matter an object has to its volume. Buoyancy is the ability or tendency to float in water, air, or some other fluid. An object with a higher density than the fluid it is in will sink. An object with a lower density than the fluid will float.
- If a **solid** object is placed in a liquid like water, it will sink or float depending on its density. For example, a beach ball is **buoyant**, or able to float, in a pool because the ball is less dense than the water.
- In our experiment, the egg sank through the layer of plain water and then floated when it reached the layer of salt water. The egg was not buoyant in the plain water, but it was buoyant in the salt water. This is because the egg was denser than the plain water but less dense than the salt water.

GLOSSARY

Buoyancy: The ability or tendency to float in water, air, or some other fluid.

Buoyant: Able to float.

Dense: Having parts that are close together.

Density: A measurement that compares the amount of matter an object has to its volume.

Float: To rest on top of a liquid.

Hypothesis: An educated guess; what we think will happen and why.

Liquid: Something that is able to flow freely like water; not a solid or a gas.

To observe: To watch something carefully.

Procedure: The steps to complete the experiment.

Question: What we want to know.

Solid: Firm or hard; not having the form of a gas or liquid.

NEW IDEAS

How could we change this experiment?

- Could we use vegetable or olive oil in place of salt water?
- Can you think of another object that would sink in ordinary water and float in salt water?

Can you think of any ideas to improve the experiment?

- Experiment by testing different amounts of salt to see at what point the egg will float (for example, 2 table-spoons and 4 tablespoons). Then, record the results.

JOURNAL

Made in the USA
Monee, IL
05 November 2026

67386153R00017